How to deal with

SBS

SICK BUILDING SYNDROME

Guidance for employers, building owners and building managers

HSE BOOKS

© Crown copyright 1995

Applications for reproduction should be addressed to HMSO

First published 1995

ISBN 0 7176 0861 1

CONTENTS

SE

SICK BUILDIN

iv

INTRODUCTION

1   Does your workforce regularly complain about irritating symptoms such as headaches, runny noses and itching? Have your supervisors noticed that people are taking much more sick leave, performing badly, or are unhappy with their surroundings? If so, it could be a case of Sick Building Syndrome. The problem could be with the building, the working environment or the way in which work is organised.

2   This booklet explains what Sick Building Syndrome is and what we know about the symptoms and the possible causes. It gives general advice on how to prevent it and how, if the symptoms occur, you can systematically investigate the problem and identify reasonably practicable improvements. The advice is aimed primarily at employers, building owners and building managers; but building designers, planners, architects, engineers, furnishers and suppliers may also find it useful.

Part 1 explains how to identify and investigate the problem.

Part 2 gives more detailed advice on how to create a good work environment.

PART 1
IDENTIFICATION
AND INVESTIGATION

What is Sick Building Syndrome?

3   Almost everyone occasionally feels unwell because they are suffering from one or more common symptoms of discomfort such as headaches, dry throat or sore eyes. But there are occasions when, for no obvious reasons, people working in particular buildings experience these sorts of symptoms more often than is usual. The symptoms tend to increase in severity with time spent in the building and improve over time or disappear away from the building. This is often described as Sick Building Syndrome.

4   The main symptoms associated with Sick Building Syndrome are:

dry or itchy skin or skin rash;

headaches, lethargy, irritability, or poor concentration.

dry or itchy eyes, nose or throat;

stuffy or runny nose;

5   The symptoms are often mild and do not appear to cause any lasting damage. To those suffering, however, they are not trivial and can cause considerable distress. In severe cases, they can affect attitudes to work and may represent a significant cost to business in the form of:

◆   reduced staff efficiency;

◆   increased absenteeism and staff turnover;

◆   extended breaks and reduced overtime;

◆   lost time complaining and dealing with complaints.

6   Sick Building Syndrome is not a recognised illness. It is simply a convenient term to describe a particular phenomenon and cannot be diagnosed precisely. It should not be confused with specific illnesses that can be directly associated with workplaces, such as humidifier fever, legionnaire's disease, the effects of exposure to specific toxic substances in the workplace or to long-term cumulative hazards such as asbestos and radon. It does not cover discomfort from adverse physical conditions in the workplace such as excessive noise, heat or cold.

What causes Sick Building Syndrome?

7   Despite extensive research we do not know the cause of Sick Building Syndrome. However, we do know that it is likely to be due to a combination of factors, the relative importance of which will be different in each case. Broadly, these factors fall into two categories:

◆   Physical or environmental factors - covering physical conditions, eg ventilation, cleaning and maintenance, and workstation layout;

◆   Job factors - such as the variety and interest of particular jobs and people's ability to control certain aspects of their work and working environment.

Poorly designed and maintained workstation

Poorly planned office organisation

8 The following list gives more detail on these factors which are identified in a review of the relevant research published by HSE in 1992[1].

Building and office design

◆ Deep plan or open plan offices of more than about ten workstations

◆ Large areas of soft furnishing, open shelving and filing

◆ New furniture, carpets and painted surfaces

Building services and maintenance

◆ Air conditioning

◆ Lighting (particularly the type and positioning which cause high glare and flicker)

◆ Low level of user control over ventilation, heating and lighting

◆ Poor design and maintenance of building services

◆ Poor standards of general repair

◆ Insufficient or badly organised office cleaning services

Indoor environment and air quality

◆ High temperature or excessive variations in temperature during the day

◆ Very low or high humidity

◆ Chemical pollutants, eg tobacco smoke, ozone, volatile organic compounds from building materials and furnishings

◆ Dust particles and fibres in the atmosphere

Job factors

◆ Routine clerical work

◆ Work with display screen equipment

9   Many of these factors are interrelated. For example, badly designed and poorly maintained air conditioning systems can create problems with ventilation and with temperature and humidity control; new furnishings can release chemical pollutants into the atmosphere; insufficient or badly organised cleaning services can create or intensify problems with dust.

10   These physical and environmental problems can be exacerbated by organisational factors. For example, lack of personal control over working conditions in open plan offices, or lack of work variation, can reduce decision-making powers and hence job satisfaction.

11   It is important to realise that not all of the risk factors occur in each case; nor do symptoms necessarily follow where the factors are present. Factors probably combine in different ways in particular cases to create the kind of environment and working arrangements in which Sick Building Syndrome occurs.

Who is affected?

12   Almost any worker can be affected by Sick Building Syndrome. Cases have occurred with many different types of building including hospitals and even in the home. Most reports, however, concern those employed in large office buildings. For this reason, Sick Building Syndrome is largely associated with office work.

13   However, not everyone is equally at risk. Workers most commonly reporting the symptoms tend to be those who have little control over their working environment and are employed in routine jobs such as general clerical work and work using display screen equipment. Women seem more at risk than men, though this partly reflects the fact that more women are employed in these areas.

How can planning prevent Sick Building Syndrome?

14   Many of the factors associated with Sick Building Syndrome relate to building and building services design. In many cases it will be very difficult, if not impossible, to change things when building and installation work have been completed. In some cases alterations may be possible but would be prohibitively expensive to carry out. The prevention of Sick Building Syndrome, therefore, needs to be tackled at an early stage during the planning of new building work, refurbishment or change of use.

15   Two broad objectives to aim for in planning are:

◆   to comply with published standards (including the Building Regulations 1991[11] and those detailed in the Chartered Institute of Building Services Engineers (CIBSE) Guides[13]); and

◆   to direct effort cost-effectively towards the best possible working environment.

16   These aims need to be applied **systematically** in the following areas:

Building services and indoor environment

 Air quality, including ventilation, outdoor air supply and air movement

 Temperature

 Humidity

 Lighting

 Noise

 Office equipment and furnishings

Maintenance

Maintenance of the building and building services systems

Cleaning operations, including office furnishings

Job factors

Management systems

Work organisation, including display screen equipment work

17   Part 2 gives more detailed guidance on the approach to take and various specific actions that may be appropriate. It also includes, where relevant, suggestions about the standards to meet.

18   But good planning is not enough. To protect the effectiveness of the design effort, it is vital to implement the plans rigorously. Construction, renovation, installation of equipment and services, and final commissioning of the building should all follow the design as precisely as possible. Any changes to the original plan need to be

checked to ensure the building as a whole will still perform as intended. In particular, materials should only be substituted when the consequences for the emission of pollutants have been assessed.

What should I do if I suspect Sick Building Syndrome?

19 If you start getting complaints from your workforce about the symptoms associated with Sick Building Syndrome, or your supervisors warn of reduced efficiency and staff unease, it is important that you investigate promptly and systematically. The problem may or may not be Sick Building Syndrome. Even if it is, there could be a number of unrelated causes requiring co-ordinated action across a variety of areas. A prompt response can help improve staff morale and make it easier to get at the real causes. However, a hasty and ill-considered response could involve you in a lot of wasted effort and money in making unnecessary changes.

20   Remember, your investigations will be most cost-effective if checks start with the most likely sources of the problem and you take the simplest actions to remedy faults as they emerge. More costly systems reviews and sophisticated remedial actions should only be considered if the simple approach does not work. You should discuss your approach with your staff or their representatives, for example the safety representative or the health and safety committee.

How do I investigate Sick Building Syndrome?

21   A reasonable order of priorities for investigation would be:

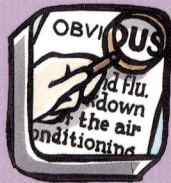

Look for the obvious

Check the symptoms

Ask the staff what the problems are

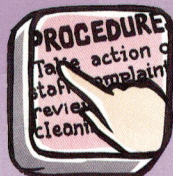

Check procedures and working practices

If these fail

Seek professional help

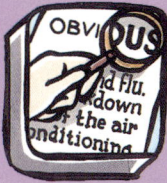

Look for the obvious

22 A sudden increase in complaints after long occupation of a building may well point to an obvious explanation. Before you do anything else just think whether there is something you know of which could be relevant, such as a local epidemic of colds or 'flu', or a breakdown in the air conditioning system. If you can, take prompt remedial action and keep the staff advised of developments.

Check the symptoms

23 If there is nothing obvious you will need to investigate the symptoms more carefully. Sick Building Syndrome symptoms are common in the community at large and the number of complaints may simply reflect what is normal in your area. On the other hand, they might also reflect the early stages of a more serious problem.

24 An analysis of staff sickness and absence records could help you to highlight any major problems, for example by pinpointing where in the building the problem might exist, or providing evidence of a viral infection. If there does seem to be a specific medical problem you will need to get a doctor or occupational health nurse to interview the sufferers and advise you on possible causes and remedies.

25   In the main, however, Sick Building Syndrome symptoms do not lead to time off work. So the best way to check the symptoms will probably be to ask the staff themselves the following questions:

◆   What are the symptoms?

◆   How frequently, and at what time of day, do they appear?

◆   How long have they been going on?

◆   Do they go away after leaving the building?

26   A group of experts working with the Royal Society of Health has designed a questionnaire specifically for this purpose. It could enable you to identify whether or not your problem is Sick Building Syndrome and give information on parts of the building, groups of staff or time of day particularly affected. The questionnaire and notes on how to use it and interpret the results will be available as a Building Research Establishment publication from the beginning of July 1995. Copies may be obtained from Construction Research Communications Ltd (address on page 30).

Ask the staff what the problems are

27   If your investigation of the symptoms does not point to the cause of your problem, you will need to look more closely at the workplace environment. The easiest way to highlight problems is, again, to ask the staff themselves. They will generally be the first to know about problems with temperature control, lighting, noise levels, stuffiness, fumes and tobacco smoke. You may also learn something about people's attitudes to their work. These can also be important clues in the search for the causes of Sick Building Syndrome.

28 For more detailed information about the situation, you can combine canvassing staff opinions with your survey of symptoms or with a more wide-ranging attitude survey. The important thing, however, is to encourage staff to make their views and complaints known by assuring them that action will be taken to alleviate problems where it is reasonably practicable to do so.

29 But remember, complaints may tell you more about the personal preferences of those putting forward the complaints than point to real problems with the working environment. For example, one person's "draught" will be another person's "fresh air". Survey results need to be carefully interpreted if they are to help you get at the real problem.

 **Check procedures and working practices**

30 If action on specific staff complaints does not alleviate the problem, you will need to carry out a comprehensive review of the building services, maintenance and cleaning procedures. Where appropriate, you should compare these with the specifications drawn up when the building was commissioned or when equipment was installed, and remedy discrepancies. In most cases, this work should be within the capabilities of your own specialist building services personnel or contractors.

31 You may also need to review other relevant areas related to the working environment such as work organisation. The guidance given in Part 2 will help you tackle these reviews systematically.

Seek professional help

32   If, in spite of all your efforts, symptoms persist, you may need to call in expert professional advice.

◆   **Building service engineers** will be able to assess the performance of the building services, particularly their ability to cope with the demand produced by the occupants and the work activities.

◆   **Occupational health doctors or nurses** can examine affected workers to identify whether they have been exposed to irritant, allergenic or toxic substances.

◆   **Occupational hygienists** will be able to assess likely sources of such exposure, and to measure the indoor environment (eg temperature, humidity and rate of air movement). Sampling and monitoring indoor air for pollutants is normally only of value in certain situations, for example if there is a known source of a particular compound or to identify the source of a suspected pollutant. Pollutants in offices are usually present at very low levels and there is little information on what levels are acceptable.

◆   **Ergonomists** can advise on job and workstation design.

◆   **Management specialists** will be able to advise on organisational factors affecting morale and job satisfaction.

But won't all this be costly?

33   Dealing with Sick Building Syndrome need not be costly if you start with the simplest things first and only move on to the more costly options if the simple things do not work. If you have to introduce major changes, ensure that you *don't make matters worse*.

34   Sophisticated remedial work, such as improving ventilation systems, changing open plan to individual offices or introducing air conditioning, can be expensive as well as very disruptive. The immediate effect can be to increase complaints by introducing new problems, eg a dusty and noisy atmosphere or disruption of workplace routine.

35   This approach can also create new long-term problems. For example, putting up new partitions can give more privacy but can also cause areas of stagnant air by interfering with air flow patterns, heat build-up from office equipment, and the release of pollutants from new furnishings. Alternatively, cleaning ductwork can release dust into the building. Each action needs to be carefully assessed for its overall impact on the workplace environment.

36   To be cost-effective remedial action will need to strike a balance between the cost of any change and the effect the change is likely to have in reducing symptoms. *Straightforward actions which can be carried out at reasonable cost and effort should be given priority.*

PART 2: CREATING A GOOD WORK ENVIRONMENT

37   This section provides more detailed advice on how to minimise risk in the main problem areas associated with Sick Building Syndrome and to help establish a reasonable working environment. It is aimed at Building Managers in particular.

38   In general, the standards quoted are those relating to good practice as set out in Health and Safety Executive guidance and other authoritative publications listed in Appendix 1. In addition, there are legal obligations imposed by the Workplace (Health and Safety and Welfare) Regulations 1992 or earlier provisions, most notably those contained in the Offices, Shops and Railway Premises Act 1963. Guidance on how to comply with the Regulations is given in the HSE publication *Workplace health, safety and welfare*: Approved Code of Practice[2].

Building services and indoor environment

Air quality, including ventilation, outdoor air supply and air movement

39   The ventilation system should deliver air of suitable quality and in sufficient quantity to:

♦   create and maintain a healthy and comfortable environment, ie provide fresh air;

♦   dilute and remove airborne impurities and pollutants, eg odours, tobacco smoke, fumes and dusts;

◆ create and maintain a comfortable temperature and humidity;

◆ prevent stagnation and draughts.

40   In offices where natural ventilation is not an option, a mechanical ventilation system will normally be adequate if it conforms with the following standards:

◆ There should be a minimum fresh-air flow of 8 litres per second per person in no smoking areas and a flow rate of up to 32 litres per second per person where heavy tobacco smoking may occur.

◆ An area with an air flow velocity in excess of 0.25 to 0.35 metres per second would be considered as draughty and of less than 0.1 metres per second as stagnant. Unless temperatures are extreme, air velocities should normally be in the region of 0.1 to 0.15 metres per second and up to 0.25 metres per second during the summer.

◆ Rooms housing office machinery such as photocopiers, and rest rooms where tobacco smoking is allowed, should have separate extract ventilation systems.

◆ Air inlets for the ventilation system should be sited to avoid introducing pollution from outside the building.

41   If you need to check that standards are being achieved, you should arrange to measure:

◆ fresh air supply rates;

◆ air velocities in the workplace.

42   But remember, turning up the ventilation rate is not the only way to deal with a problem with air quality. A more effective response will be to identify what is polluting the air, whether it be odours, tobacco smoke, fumes or dusts, and tackle the problem at source by reducing the amount of pollutants being released into the office environment. This may require carrying out a limited monitoring programme to identify the pollutants and then to assess the effectiveness of the remedial measures taken.

43   More detailed guidance on ventilation is given in HSE leaflets *Ventilation of the workplace*[3] and *Measurement of air change rates in factories and offices*[4]. Advice on smoking in the workplace is given in HSE leaflet *Passive smoking at work*[9]. Other sources of guidance are the CIBSE Guide A Section A4: *Air Infiltration and Natural Ventilation*[12] and BSRIA publication *Ventilation effectiveness in mechanical ventilation systems*[15]. Outdoor air supply rates are given in the CIBSE Guide B Section 2B: *Ventilation and Air Conditioning Requirements*[12].

Temperature

44   Failure to control the workplace temperature is unlikely by itself to cause Sick Building Syndrome. But excessive temperatures and wide variations in temperature can influence other factors by, for instance, increasing the possibility of exposure to airborne pollutants.

45   There is no defined "ideal" standard workplace temperature. The recommended minimum for workrooms in general is 16°C but actual need will vary according to circumstances and the particular types of work involved. In the normal office environment, it would be reasonable to maintain a temperature of around 19°C. To check that this standard is being achieved, you should measure the air temperature with an ordinary dry bulb thermometer, close to workstations, at working height and away from windows. In considering systems for maintaining the overall heating standards, you will need to take full account of any localised effects of sunlight and radiant heat from office machinery.

46   Remember, the heating or cooling system used should not allow dangerous or offensive fumes to escape into the workplace.

Humidity

47   As with temperature, humidity on its own is unlikely to be a cause of Sick Building Syndrome. Unreasonable humidity levels in combination with other factors can, however, exacerbate problems. For example, high humidity encourages the growth of harmful bacteria; low humidity contributes to a dusty atmosphere and to dry eyes, nose, throat and skin.

48   In offices, it is generally considered in the interests of worker efficiency that humidity be maintained in the range of 40% to 70%. In warm offices, the relative humidity should be at the lower end of this range. Controls should be checked frequently. The checks should also cover the operation and cleanliness of the humidifying equipment.

Lighting

49   In offices, the lighting should:

◆   where possible, be designed to give individual control;

◆   utilise natural light;

◆   avoid glare, flicker, and noise;

◆   be kept clean and defective units replaced promptly;

◆   be appropriate for the work, in particular for work with display screen equipment.

50   More detailed guidance on lighting is given in the HSE leaflet *Lighting at work*[5] and the CIBSE publication *Code for interior lighting*[14].

Noise

51   The intensity of noise is unlikely to cause Sick Building Syndrome symptoms on its own. More probably it will be the disruptive effect on the workforce who may perceive the noise as an imposed, unnecessary nuisance.

52   Office equipment which has low noise emission characteristics will help reduce the noise burden in the office. However, there are other sources of unpredictable and intrusive noise which may need careful consideration, eg air at outlet vents and in ductwork, water in pipes, and vibration from air conditioning plant. Where there are complaints of noise from office equipment or building services systems, expert advice may be required.

53   Because of the way noise is absorbed or reflected by surrounding surfaces, it is particularly important that any renovations which involve the removal of carpets, soft furnishings or false ceilings take full account of the effect on overall noise levels.

Office equipment and furnishings

54   Office equipment and new furnishings can release chemicals, known as volatile organic compounds, into the workplace, and these have been associated with Sick Building Syndrome. Most problems should be solved during commissioning or the first months of occupation but you may find that symptoms are associated with the introduction of new equipment and furnishings. If this is the case you may need to consult the experts to determine the best course of action.

55   The difficulty in analysing the problem is that individual pollutants are often at the limits of detection of available

methods and at levels several times lower than current occupational exposure limits. Sampling can be complex and the use of certain sampling techniques may be expensive. This means that local testing for airborne pollutants is an effective option only in a limited number of situations.

56   However, more information on the emission characteristics of building and furnishing materials and office equipment is becoming available. For new projects and refurbishments, this information will be an important guide to help with the selection of new or replacement materials. Where possible, you should give preference to low emission products, eg furniture with low formaldehyde emissions.

Maintenance

Maintenance of the building and the building services systems

57   Good maintenance procedures are often the best way to prevent or reduce Sick Building Syndrome symptoms, and careful planning will help produce the best results. Make sure your maintenance scheme covers:

◆   the fabric of the building;

◆   building services (eg heating, ventilation, air-conditioning and lighting systems);

◆   building furnishings;

◆   office equipment.

58   An effective scheme will include drawing up schedules to record the type and frequency of:

◆   system performance testing;

◆   visual inspections of physical condition;

- examination of system components;

- replacement of items with fixed life spans, such as filters;

- cleaning operations.

59   Other factors you will need to consider in designing an effective scheme include:

- facilities available for cleaning and specialist testing;

- practical problems in carrying out necessary tasks including difficulty of access to ventilation ducting;

- changing requirements during the year, eg the control settings of the building services (ie the heating, ventilation, air-conditioning and lighting systems) need to be adjusted regularly as outside conditions change.

60   In addition, a record of your cleaning, inspection, testing and maintenance operations will help to sustain efficient operating systems. Laying down procedures for bringing abnormal conditions to the attention of the building management will help you spot and tackle problems promptly.

61   An integral part of the planning stage will be to identify those responsible for implementing and supervising the various provisions of the scheme. This will involve making arrangements for them to be suitably trained and equipped to carry out their duties effectively, including the ability to monitor maintenance operations carried out by maintenance contractors. To avoid confusion and ensure that everyone continues to be fully aware of their part in the system, it would be helpful to have the scheme written down and made available for reference purposes.

62   More detailed guidance on building maintenance can be found in British Standard BS8210 *Guide to Building Maintenance Management*[11] and CIBSE's *Maintenance Management for Building Services*[13].

Cleaning operations, including office furnishings

63   Cleaning can be a major factor in preventing Sick Building Syndrome. Cleaning patterns for particular areas should be set according to individual circumstances but the following frequencies of cleaning operations are suggested as a guide:

◆   wet areas of plant including cooling coils and humidifiers (annually);

◆   ventilation systems including grills and vents (annually);

◆   windows and light fittings (monthly/3 monthly);

◆   internal surfaces, office carpeting, furnishings and furniture including desks and chairs (daily), and deep cleaning of soft furnishings (annually).

64   Cleaning methods and choice of cleaning materials are also important. Where plant and equipment are involved, the manufacturer's instructions should be followed. In general, avoid cleaning materials that give off strong odours. Avoid, too, methods which might disturb surfaces and raise dusts, fibres and microbiological particles into the air. Vacuum cleaners fitted with high efficiency final filters can help achieve this.

65   The effectiveness of cleaning operations can be reduced where desks and surrounding areas are obstructed with papers and other clutter. If appropriate, therefore, you could consider introducing a clear desk and clear floor policy to apply at the end of the working day.

66   In addition, the timing of workplace cleaning should be considered carefully. For instance, cleaning operations carried out in the evening rather than in the early morning will help prevent odours from ash trays and waste bins being absorbed into the office furnishings which can cause a continuing problem the following day. It will also provide time for dust to settle and ensure a better air quality at the start of the new working day.

Job factors

Management systems

67   Although there is little direct evidence that management and work organisation systems can cause Sick Building Syndrome, it is very likely that dissatisfaction with the job and working conditions can exacerbate symptoms and lead to more complaints. Symptoms are more likely to be reported among those who have least control over their working environment. Typical problem areas are large, open plan offices where staff have little privacy, limited opportunity to alter heating, ventilation or lighting, and little say in how their sometimes monotonous work is organised.

68   Well motivated staff, confident that their concerns are taken seriously, will be more likely to give early warning of developing problems and more appreciative of efforts to improve the situation, particularly where scope for altering the physical characteristics of office areas may be limited.

69   Good communications and good relationships between management and staff will be important ingredients in helping to keep management in touch with staff concerns and staff informed about management's efforts to deal with problems. Involving workers representatives or the safety committee, where one has been established, can be a useful way to help achieve this.

**Work organisation,
including display screen equipment work**

70   Dissatisfaction can result from bad job and workstation design. Proper job design is probably the most important aspect of work for most people. If you can, involve individual members of staff in designing their own jobs and setting their own targets. Take full advantage of opportunities to break up routine procedures by, for example, introducing some variety in the tasks set, well timed rest breaks and regular job rotation.

71   Design also plays an important part in influencing staff attitudes to open plan, particularly where the areas are large. Open plan access to external windows where possible, the provision of plants, and careful choice of colour scheme can all help towards greater satisfaction.

72   Care should be taken to ensure workstations are suitable for the people using them, reviewed regularly and adapted, as necessary, when there is a change of staff.

73   For detailed advice on the safe design and use of workstations for display screen equipment see HSE publications *Display screen equipment work - Guidance on the Health and Safety (Display Screen Equipment) Regulations 1992*[6], *VDUs: an easy guide to the Regulations*[7] and *Working with VDUs*[8].

Where can I obtain further information?

74   The Environmental Health Services of your local authority or the local HSE area office will be able to give you general advice on Sick Building Syndrome and how to deal with it. They may also be able to put you in touch with local specialist organisations who can offer you professional advice. A list of organisations with interests in the indoor environment and details of relevant publications are given in the Appendix.

APPENDIX: RELEVANT ORGANISATIONS and PUBLICATIONS

Organisations which can provide further information

Area offices of the Health and Safety Executive
(for addresses see local phone book)

Environmental Health Services of your local authority
(for addresses see local phone book)

The Department of the Environment
Romney House, 43 Marsham Street
London SW1P 3PY
Tel: 0171 276 8000

The Building Research Establishment
Garston
Watford WD2 7JR
Tel: 01923 894040

Construction Research Communications Ltd
33-39 Bowling Green Lane
London EC1R 0DA
Tel: 0171 837 1212

The Chartered Institution of Building Services Engineers
Delta House, 222 Balham High Road
London SW12
Tel: 0181 675 5211

Publications issued by HSE

On sick building syndrome

1   G J Raw *Sick Building Syndrome: a review of the evidence on causes and solutions* HSE Contract Research Report No.42/1992 HSE Books ISBN 0 11 886364 9

On employer duties in the workplace

2   *Workplace Health, Safety and Welfare: Workplace (Health, Safety and Welfare) Regulations 1992*: Approved Code of Practice and Guidance (L24) HSE Books ISBN 0 11 886333 9

3   *Ventilation of the workplace* EH22(Rev) HSE Books 1988 ISBN 0 11 885403 8

4   *Measurement of air change rates in factories and offices* (MDHS73) HSE Books 1992 ISBN 0 11 885693 6

5   *Lighting at work* HS(G)38 HSE Books 1987 ISBN 0 11 883964 0

6   *Display Screen Equipment Work - Guidance on the Health and Safety (Display Screen Equipment) Regulations 1992* (L26) HSE Books 1992 ISBN 0 7176 0410 1

7   *VDUs: an easy guide to the Regulations*   HS(G)90 HSE Books 1994 ISBN 0 7176 0735 6

8   *Working with VDUs* IND(G)36L(Rev) 1992 (free booklet, available in priced packs)

9   *Passive smoking at work* IND(G)63L(Rev)1992 (free booklet, available in priced packs)

Other publications on standards

Regulations

10   *The Building Regulations* 1991 (SI 1991/2768) HMSO
ISBN 0 11 015887 3

British Standards Institution, London

11   BS 8210:1986 *Guide to Building Maintenance
Management: Recommendations for a Systematic
Approach in the United Kingdom* ISBN 0 580 15241 3

The Chartered Institute of Building Services Engineers (CIBSE), London

12   *The CIBSE Guides (1986) - Volume A: Design Data*
ISBN 0 900953 29 2, *Volume B: Installation and Equipment
Data* ISBN 0 900953 30 6, and *Volume C: Reference Data*
ISBN 0 90053 31 4

13   *Maintenance Management for Building Services* 1994
TM17 ISBN 0 900953 683

14   *Code for interior lighting* 1994 ISBN 0 900953 64 0

The Building Services Research and Information Association (available through CIBSE)

15   *Ventilation effectiveness in mechanical ventilation
systems* (1988) BTN01/88 ISBN 0 860221 89 X

Printed and published by
the Health and Safety Executive C80   03/95